CAREERS IN SCIENCE

What Do METEOROLOGISTS Do?

BENJAMIN PROUDFIT

PowerKiDS press

NEW YORK

Published in 2022 by The Rosen Publishing Group, Inc.
29 East 21st Street, New York, NY 10010

First Edition

Portions of this work were originally authored by Kristen Rajczak and published as *Be a Meteorologist!* All new material in this edition was authored by Benjamin Proudfit.

Editor: Greg Roza
Book Design: Michael Flynn

Photo Credits: Cover Matthew Peyton/Getty Images; (background texture) Anna Timoshenko/Shutterstock.com; pp. 5 (meteorologist), 11, 13, 16 Joe Raedle/Getty Images; p. 5 (Aristotle) Photo Josse/Leemage/Corbis Historical/Getty Images; p. 7 courtesy of the Library of Congress; p. 8 simonkr/E+/Getty Images; p. 9 Portland Press Herald/Getty Images; p. 14 Michael Cohen/Getty Images; p. 15 Sam Greenwood/Getty Images; p. 17 Streeter Lecka/Getty Images; p. 19 Marcus Yam/Los Angeles Times/Getty Images; p. 20 pikepicture/iStock/Getty Images; p. 21 Simon Phelps Photography/Moment/Getty Images; p. 23 (NEXRAD) Andrei Stanescu/iStock/Getty Images; p. 23 (hurrican drone) Saul Loeb/AFP/Getty Images; p. 25 Images By Kenny/Shutterstock.com; p. 26 Photoagriculture/Shutterstock.com; p. 27 RaksyBH/Shutterstock.com; p. 29 Jeff Greenberg/Universal Images Group/Getty Images.

Library of Congress Cataloging-in-Publication Data

Names: Proudfit, Benjamin, author.
Title: What do meteorologists do? / Benjamin Proudfit.
Description: New York : PowerKids Press, [2022] | Series: Careers in science | Includes index.
Identifiers: LCCN 2020040288 | ISBN 9781725329591 (library binding) | ISBN 9781725329577 (paperback) | ISBN 9781725329584 (6 pack)
Subjects: LCSH: Meteorologists--Juvenile literature. | meteorology--Vocational guidance--Juvenile literature.
Classification: LCC QC869.5 .P76 2022 | DDC 551.5023--dc23
LC record available at https://lccn.loc.gov/2020040288

Manufactured in the United States of America

CPSIA Compliance Information: Batch #CSPK22. For further information contact Rosen Publishing, New York, New York at 1-800-237-9932.

CONTENTS

CHECK THE FORECAST

Whether you're planning a trip that's a few months away or a picnic this weekend, you're likely to check what the weather may be like. This is called the weather forecast, and it's created by scientists called meteorologists. The science of meteorology looks at the causes of and changes in weather and **climate**. Meteorologists are scientists who study Earth's **atmosphere**. They're also concerned with how people's activities affect Earth's atmosphere.

Forecasting the weather is just one part of a meteorologist's job. Meteorologists watch how weather and climate conditions change in a place over time. They also study how people's activities, such as pollution, can affect Earth's atmosphere.

Meteorologists use knowledge they gather through observation and study, plus past experience, to guess what could happen with weather and climate in the future.

WHAT'S THE WORD?

The term "meteorology" comes from a Greek word meaning "something that happens high in the sky." The ancient Greeks tried to understand how clouds, wind, and rain worked together. Aristotle, a great thinker in ancient Greece, wrote a book called *Meteorologica* more than 2,000 years ago. It included his studies of Earth's atmosphere.

Aristotle

PAST WEATHER STUDY

The weather has been studied for thousands of years! By the 1600s, scientists had already begun to record wind direction and rainfall measurements. The thermometer was invented around 1600. By the 1800s, even more meteorological tools had been invented to help scientists make sense of the weather around them. They began to understand that weather moves in systems across Earth, and this allowed them to do limited forecasting. When the **telegraph** was introduced, weather reports could be sent anywhere instantly.

Computers were first used to follow atmospheric changes in the 1950s. During the 1960s, the first **satellites** used for weather observation were sent into **orbit**.

Weather balloons were commonly used to gather atmospheric information in the first half of the 20th century.

WORLD WARS

Military operations during World War I (1914–1919) and World War II (1939–1945) often depended on the weather. It was for this reason that great advances in meteorological **technology** occurred during both wars. The military trained scientists and did meteorological **research** in order to better **predict** the weather. **Radar** was the biggest achievement made in meteorology technology. It's still used today!

METEOROLOGISTS ON TV OR RADIO

Broadcast meteorologists are some of the most recognizable scientists to the public. They work at TV and radio stations explaining the weather forecast for others to understand. These scientists need to know a lot about weather as well as be good at **communicating** their knowledge to others.

Broadcast meteorologists may help make the graphics, or computer-made pictures, used to show how the weather moves across the weather map you see on TV!

WORKING TOGETHER

All kinds of meteorologists look to other scientists, including other meteorologists, around the world when predicting the weather. That's because weather systems in one place—even one far away—can affect the weather in another place. When you see the weather forecast on TV, it's likely a product of many people around the world observing the weather!

Not all people reporting the weather on TV or the radio are meteorologists. Weathercasters may simply be people who pass on weather reports received from forecasting services, like the National Weather Service. Broadcast meteorologists, on the other hand, follow the weather and use facts and figures, or data, to create the forecast for their area using their knowledge and observations. Broadcast meteorologists are especially needed because they're the ones who warn the public about extreme weather, such as tornadoes, snowstorms, or hurricanes.

RESEARCH IT!

Though they may be the most visible kinds of meteorologists, broadcast meteorologists are just one kind of these scientists! Research meteorologists spend their time gathering measurements and other data about their area of research in order to learn more and make new discoveries.

One type of research meteorologist is a climatologist. Climatologists study climate using weather records from past years. They look for changes over time and predict future **trends**. Archive meteorologists also look at past weather records to make sure they're as accurate and detailed as possible. Some meteorologists go to school for engineering. They may become instrument specialists. These scientists design new tools and systems for recording the weather.

Research meteorologists may study how the atmosphere has changed over time, how hurricanes form, how animals and the atmosphere affect each other, and more!

FORENSIC METEOROLOGISTS

Could knowing the weather of a certain day help solve a crime? Sometimes forensic meteorologists are called in by the police or others working on a case. They put together what the weather and conditions were like at a certain day or time. Knowing whether there was lightning or whether the sun was in a driver's eyes could change the outcome of a case.

WHERE TO WORK

Many meteorologists in the United States work for the U.S. government. They often work for the National Oceanic and Atmospheric Administration (NOAA). The National Weather Service is part of the NOAA. Its mission is to "provide weather, water, and climate data, forecasts and warnings for the protection of life and property and enhancement of the national economy."

Other government meteorologists may work for the National Aeronautics and Space Administration (NASA). Over time, NASA meteorologists have discovered that even slight cloud cover can affect how a rocket will take off. This can make their job tough. They need to know what the weather will be like at an exact time and location, not just generally over a few hours in an area.

MILITARY METEOROLOGISTS

Meteorologists are also a very important part of the U.S. military. Knowing the weather and sea conditions and how they might change is a central part of planning military missions. Meteorologists are stationed at U.S. military bases all over the world. They even work on ships that travel around the world at sea!

On May 30, 2020, the SpaceX Falcon 9 rocket launched from Cape Canaveral, Florida. It would have launched three days before, but the weather conditions weren't right.

JOB OUTLOOK

Meteorology is a very exciting area of study. But there are more meteorologists entering the field each year than there are jobs. It's important to know about the many different kinds of places a meteorologist, or someone with that kind of knowledge, can work beyond NOAA or NASA. Airlines, for example, may hire teams of meteorologists to help with flight plans or forecasting weather around an airport.

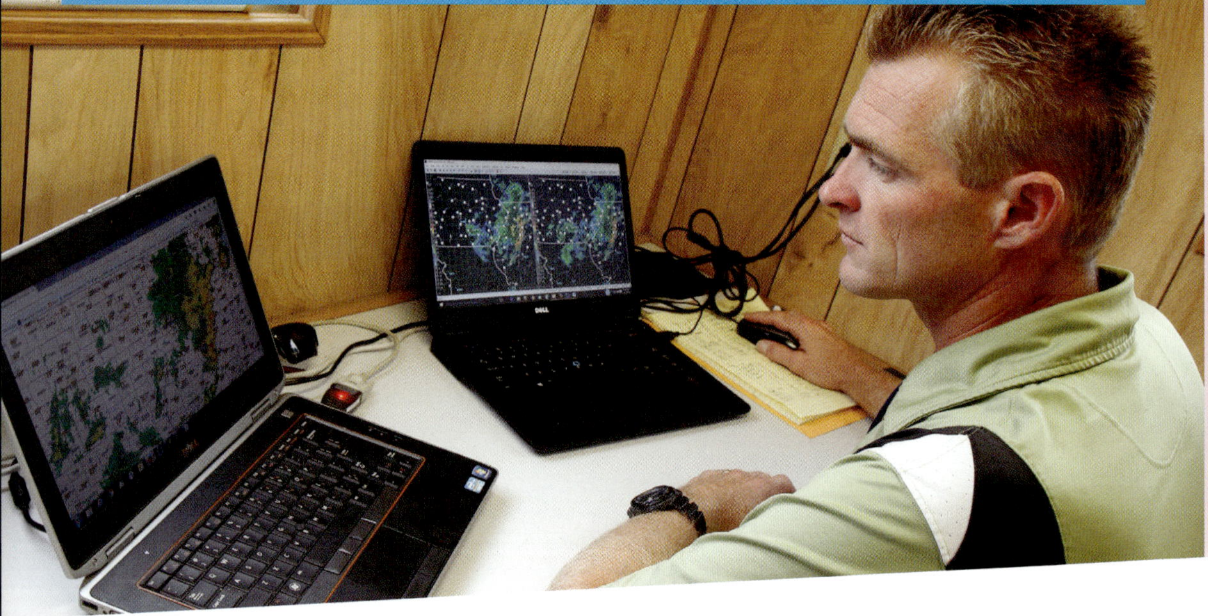

There are many companies that need to know the weather to do their work well. Construction businesses, transportation companies, and fishermen all might use meteorologists' knowledge to prepare for changing conditions. If a business is large enough, it may employ its own meteorologist. Others use private forecasting companies that hire meteorologists to forecast the weather and then sell that knowledge to companies and people that want it.

Professional sports teams also use meteorologists to know how weather may affect upcoming games. The New York Jets and Philadelphia Eagles get their weather reports from the meteorologists at a company called WeatherWorks.

After working as a scientist or forecaster for many years, a meteorologist may teach at a college or high school.

Big games and tournaments use meteorologists to track the weather in order to keep players and fans safe.

DAILY WORK

The cool thing about being a meteorologist is that every day is different—because the weather is! Whether they're working for the government or a private company, meteorologists commonly work in weather stations and observation centers. They collect data and satellite reports to make short-term and long-term forecasts.

National Weather Service Weather Forecast Offices have meteorologists on duty 24 hours a day every day of the year. That's because the weather doesn't stop just because it's night in one part of the world. There's at least one short-term forecaster who's working on the weather for the next one to two days. A long-term forecaster is also on duty looking further ahead.

Because meteorologists track big storms and weather events, people can find out how to be safe where they live.

WATCHING AND WARNING

Meteorologists are always looking out for the possibility that a big storm, such as a hurricane, is forming. They pass important information to the public through state and city officials and TV and radio reports. The National Weather Service issues "watches" if the conditions are right for extreme weather, such as thunderstorms that could cause power outages or floods. "Warnings" are issued when that weather event is actually occurring in an area.

FORECASTING

It takes many meteorologists across the country and around the world to create a forecast. In the last few decades, weather forecast accuracy has improved a lot. The three-day forecasts today are more accurate than one-day forecasts of 20 years ago. Meteorologists' five-day forecasts are better than three-day forecasts in 1970. Better technology and more knowledge about how the oceans and atmosphere interact play a big role in this.

Here's how a forecast comes together: First, weather observers record measurements about four times a day at weather stations on land and at sea all over the world. More than 500 weather stations release weather balloons into the atmosphere to collect data higher above Earth's surface.

Some meteorologists spend some of their time experiencing the weather they report on!

GLOBAL CHANGES

Meteorologists are some of the scientists working to learn more about global climate change, including the slow warming of the atmosphere and oceans. As these changes affect the weather, meteorologists are studying the problems climate change may cause, such as an increase in destructive storms. They're working on solutions to these problems too.

Next, data is sent to world weather centers in Russia, the United States, and Australia. Computers put the facts together to give a good picture of the global weather. Then, the National Weather Service in Washington, D.C., takes the data and applies it to the United States using computer models.

The models are sent to local weather centers. They're used to make even more focused forecasts for cities and towns. Broadcast meteorologists or weathercasters use them to give the forecast on TV or radio.

Just one weather station wouldn't be able to collect enough data to make any weather predictions! Meteorologists from all over the world work together so they're better able to predict weather that could cause harm and alert those who could be affected.

COMPUTER POWER

Meteorologists don't do forecasting all on their own. They use some of the fastest and most powerful computers in the world. Different weather conditions are entered into a computer so it can create models of weather systems. The models then make weather "happen" on the computer. This allows meteorologists to see what might happen days or weeks in the future.

From thunderstorms to tornadoes, the weather affects everyone. Meteorologists make sure you have the information you need!

TOOLS AND TECH

Understanding and using technology is an important part of being a meteorologist. One of the most important tools, radar, finds and measures rain, snow, and other precipitation. Doppler radar is a special kind of radar used to measure wind speed and direction.

There are also many weather satellites orbiting Earth. These allow meteorologists to see weather systems moving around the planet. Early weather satellites could only show pictures, but today, satellites can also monitor wind speed, temperature, and other atmospheric conditions.

Specially equipped aircraft can fly right into weather systems. Some have brave pilots flying them right into hailstorms. Others are unmanned aircraft, or drones. They can go where it's unsafe to send a person.

Next Generation Weather Radar, or NEXRAD, are found across the United States, including near San Jose, California, shown here.

HURRICANE HUNTERS

The NOAA does important research on hurricanes by sending aircraft very close to big storms! These Hurricane Hunters are several different kinds of aircraft that have special gear and a highly trained crew to use it. They measure changes in temperature, pressure, wind speed, and more to monitor how a storm is moving or building. The work of Hurricane Hunters has made hurricane prediction better, which can save lives!

hurricane drone

BECOME A METEOROLOGIST!

Like with many other careers in science, doing well in school is important. Once you enter high school, you'll need to take many math and science classes, including physics, Earth science, and an advanced math class called calculus. These classes will help you once you begin studying in college—and a college degree is something all meteorologists need.

Learning about new technology is also helpful to future meteorologists. Study how computers work with different kinds of programs to aid you on your path to a career in meteorology.

Speaking a foreign language, such as French, Russian, or German, can be a good thing too. Meteorologists often look at research from other parts of the world. They often work with people from around the world!

GET TALKING!

Since meteorologists work a lot with other scientists, it's a good idea to be able to communicate your ideas well. Taking a public speaking class could help you learn to present your research clearly. Paying attention in English and writing classes will help you when you need to share research or ideas with scientists around the world in a publication or another written way.

For those who want to be broadcast meteorologists, even more classes are needed! You'll need to learn how to communicate clearly on camera, sometimes in very stressful weather situations.

Many meteorologists have college degrees in atmospheric science, computer science, math, and physics. They choose to take classes that teach them about oceanography, or the study of the oceans, and other important topics involved in meteorology.

Before becoming a meteorologist, many students follow a working meteorologist around for a time, observing their day-to-day tasks. Students often complete an **internship** at a weather station or atmospheric research center.

Meteorologists who want to do scientific research usually need to have more than just a college degree. They need to go to graduate school to earn a master's degree or, even higher, a doctoral degree. During these studies, they'll choose an area of meteorology they want to focus on, such as climatology.

Would you like to work at the National Weather Center or another weather station? Apply for an internship! Sometimes, an internship can lead to a job after school.

USEFUL SCIENTISTS

You don't have to wait until college to start learning about meteorology! Start now by working hard in science, math, and computer classes. Spend time outside observing the world around you. Take notes about the daily weather, including the high and low temperatures and other measurable conditions.

Meteorologists are scientists who learn about weather and climate changes around the world so that we can know what to expect in the future. Their useful knowledge can also show how people are affecting Earth and its atmosphere. Because of the work of meteorologists, we know more about our planet and have found ways to better care for it together.

Talking to a working meteorologist is a great way to find out if this career in science is for you.

MEET A METEOROLOGIST

Many meteorologists work in places that bring in school groups for educational visits. If you are interested in becoming a meteorologist—or just want to know more about predicting the weather—talk to your parents or science teacher. They might be able to set up a classroom visit with a working meteorologist or field trip to a weather center near you.

GLOSSARY

atmosphere: The mixture of gases that surround a planet.

climate: The average weather conditions of an area.

communicate: To share thoughts or feelings by sound, movement, or writing.

internship: A position as an intern, or a person who works for a period time in order to gain experience, often wthout pay.

orbit: To travel in a circle or oval around something, or the path used to make that trip

predict: To guess what will happen in the future based on facts or knowledge.

radar: A machine that uses radio waves to locate and identify distant objects.

research: Studying to find something new.

satellite: An object that circles the planet in order to collect and send data to scientists on Earth.

technology: Using science, engineering, and other industries to invent useful tools or to solve problems. Also, a machine, piece of equipment, or method created by technology.

telegraph: A method of communicating using electric signals sent through wires.

trend: A general direction of change.

FOR MORE INFORMATION

BOOKS

Rajj, Emily. *Meteorologists.* North Mankato, MN: Pebble, 2020.

Schuh, Mari C. *Predicting the Weather.* Vero Beach, FL: Rourke Educational Media, 2020.

WEBSITES

Kids' Crossing
eo.ucar.edu/kids/
Learn much more about Earth's weather with facts, photographs, quizzes, and more.

SciJinks
scijinks.gov/
Check out this website that explores weather and how it's forecasted.

INDEX